Experiments in Plant Hybridisation

GREGOR MENDEL

NEW YORK

Experiments in Plant Hybridisation

Experiments in Plant Hybridisation was originally published in 1909.

For information, address:
P.O. Box 416, Old Chelsea Station
New York, NY 10011

or visit our website at:
www.cosimobooks.com

Cover Design by www.popshopstudio.com

ISBN: 978-1-60520-257-0

Experiments which in previous years were made
with ornamental plants have already afforded evidence
that the hybrids, as a rule, are not exactly intermediate
between the parental species. With some of the more
striking characters, those, for instance, which relate to
the form and size of the leaves, the pubescence of
the several parts, etc., the intermediate, indeed, is nearly
always to be seen; in other cases, however, one of the two
parental characters is so preponderant that it is difficult,
or quite impossible, to detect the other in the hybrid.

—from "4. The Forms of the Hybrid"

EXPERIMENTS IN PLANT HYBRIDISATION

Gregor Mendel

1. *INTRODUCTORY REMARKS*

Experience of artificial fertilisation, such as is effected with ornamental plants in order to obtain new variations in colour, has led to the experiments which will here be discussed. The striking regularity with which the same hybrid forms always reappeared whenever fertilisation took place between the same species induced further experiments to be undertaken, the object of which was to follow up the developments of the hybrids in their progeny.

Editor's Note. Mendel's paper, "Versuche über Pflanzenhybriden", was read at meetings of the Brunn Natural History Society on 8th February and 8th March 1865, and was published in the *Verhandlungen des Naturforschenden Vereins in Brünn*, 4, 1865, which appeared in 1866. An English translation was made by the Royal Horticultural Society of London and published in volume 26 of the Society's Journal in 1901. It is the modified version of this translation as given by W. Bateson in his book, *Mendel's Principles of Heredity* (C.U.P., 1909) that is reprinted here. The footnotes and changes due to Bateson are shown within square brackets whilst the few small changes introduced with this reprinting are enclosed within double square brackets. Attention should perhaps be drawn to one of these changes. Some confusion has arisen in the past from the use of the same word " trial " for Mendel's preliminary two-year tests and for his later experimental breeding work. Here, "trial" is used to translate "Probe" with which Mendel describes his tests with the 34 varieties of peas in 1854-55; "Versuch" and "Experiment", which Mendel uses in referring to the rest of his programme, are translated throughout as "experiment".

To this object numerous careful observers, such as Kölreuter, Gärtner, Herbert, Lecoq, Wichura and others, have devoted a part of their lives with inexhaustible perseverance. Gärtner especially in his work "Die Bastarderzeugung im Pflanzenreiche" [The Production of Hybrids in the Vegetable Kingdom], has recorded very valuable observations; and quite recently Wichura published the results of some profound investigations into the hybrids of the Willow. That, so far, no generally applicable law governing the formation and development of hybrids has been successfully formulated can hardly be wondered at by anyone who is acquainted with the extent of the task, and can appreciate the difficulties with which experiments of this class have to contend. A final decision can only be arrived at when we shall have before us the results of detailed experiments made on plants belonging to the most diverse orders.

Those who survey the work done in this department will arrive at the conviction that among all the numerous experiments made, not one has been carried out to such an extent and in such a way as to make it possible to determine the number of different forms under which the offspring of hybrids appear, or to arrange these forms with certainty according to their separate generations, or definitely to ascertain their statistical relations.*

It requires indeed some courage to undertake a labour of such far-reaching extent; this appears, however, to be the only right way by which we can finally reach the solution of a question the importance of which cannot be overestimated in connection with the history of the evolution of organic forms.

The paper now presented records the results of such a detailed experiment. This experiment was practically confined to a small plant group, and is now, after eight years' pursuit, concluded in all essentials. Whether the plan upon which the separate experiments were conducted and carried out was the best suited to attain the desired end is left to the friendly decision of the reader.

---

* [It is to the clear conception of these three primary necessities that the whole success of Mendel's work is due. So far as I know this conception was absolutely new in his day.]

2. SELECTION OF THE EXPERIMENTAL PLANTS

The value and utility of any experiment are determined by the fitness of the material to the purpose for which it is used, and thus in the case before us it cannot be immaterial what plants are subjected to experiment and in what manner such experiments are conducted.

The selection of the plant group which shall serve for experiments of this kind must be made with all possible care if it be desired to avoid from the outset every risk of questionable results.

The experimental plants must necessarily—

1. Possess constant differentiating characters.

2. The hybrids of such plants must, during the flowering period, be protected from the influence of all foreign pollen, or be easily capable of such protection.

The hybrids and their offspring should suffer no marked disturbance in their fertility in the successive generations.

Accidental impregnation by foreign pollen, if it occurred during the experiments and were not recognised, would lead to entirely erroneous conclusions. Reduced fertility or entire sterility of certain forms, such as occurs in the offspring of many hybrids, would render the experiments very difficult or entirely frustrate them. In order to discover the relations in which the hybrid forms stand towards each other and also towards their progenitors it appears to be necessary that all members of the series developed in each successive generation should be, *without exception*, subjected to observation.

At the very outset special attention was devoted to the *Leguminosae* on account of their peculiar floral structure. Experiments which were made with several members of this family led to the result that the genus *Pisum* was found to possess the necessary qualifications.

Some thoroughly distinct forms of this genus possess characters which are constant, and easily and certainly recognisable, and when their hybrids are mutually crossed they yield perfectly fertile progeny. Furthermore, a disturbance through foreign pollen cannot easily occur, since the fertilising organs are closely packed inside the keel and the anthers burst within the bud, so that the stigma becomes covered with pollen even before the flower opens. This circumstance is of especial importance. As additional advantages worth mentioning, there may be cited the easy culture of these plants in the open ground and in pots, and also their relatively short period of growth. Artificial fertilisation is certainly a somewhat elaborate process, but nearly always succeeds. For this purpose the bud is opened before it is perfectly developed, the keel is removed, and each stamen carefully extracted by means of forceps, after which the stigma can at once be dusted over with the foreign pollen.

In all, thirty-four more or less distinct varieties of Peas were obtained from several seedsmen and subjected to a two years' trial. In the case of one variety there were noticed, among a larger ⟦considerable⟧ number of plants all alike, a few forms which were markedly different. These, however, did not vary in the following year, and agreed entirely with another variety obtained from the same seedsman; the seeds were therefore doubtless merely accidentally mixed. All the other varieties yielded perfectly constant and similar offspring; at any rate, no essential difference was observed during ⟦the⟧ two trial years. For fertilisation twenty-two of these were selected and cultivated during the whole period of the experiments. They remained constant without any exception.

Their systematic classification is difficult and uncertain. If we adopt the strictest definition of a species, according to which only those individuals belong to a species which under precisely the same circumstances display precisely similar characters, no two of these varieties could be referred to one species. According to the opinion of experts, however, the majority belong to the species *Pisum sativum*; while the rest are regarded and classed, some as sub-species of *P. sativum*, and some as independent species, such as *P. quadratum*, *P. saccharatum*, and *P. umbellatum*. The positions, however, which may be assigned to them in a classificatory system are quite immaterial for the purposes of the experiments in question. It has so far been found to be just as impossible to draw a sharp line between the hybrids of species and varieties as between species and varieties themselves.

3. DIVISION AND ARRANGEMENT OF THE EXPERIMENTS

If two plants which differ constantly in one or several characters be crossed, numerous experiments have demonstrated that the common characters are transmitted unchanged to the hybrids and their progeny; but each pair of differentiating characters, on the other hand, unite in the hybrid to form a new character, which in the progeny of the hybrid is usually variable. The object of the experiment was to observe these variations in the case of each pair of differentiating characters, and to deduce the law according to which they appear in the successive generations. The experiment resolves itself therefore into just as many separate experiments as there are constantly differentiating characters presented in the experimental plants.

The various forms of Peas selected for crossing showed differences in the length and colour of the stem; in the size and form of the leaves; in the position, colour, and size of the flowers; in the length of the flower stalk; in the colour, form, and size of the pods; in the form and size of the seeds; and in the colour of the seed-coats and of the albumen [cotyledons]. Some of the characters noted do not permit of a sharp and certain separation, since the difference is of a "more or less" nature, which is often difficult to define. Such characters could not be utilised for the separate experiments; these could only be applied to characters which stand out clearly and definitely in the plants. Lastly, the result must show whether they, in their entirety, observe a regular behaviour in their hybrid unions, and whether from these facts any conclusion can be come to regarding those characters which possess a subordinate significance in the type.

The characters which were selected for experiment relate:

1. To the *difference in the form of the ripe seeds*. These are either round or roundish, the depressions, if any, occur on the surface, being always only shallow; or they are irregularly angular and deeply wrinkled (*P. quadratum*).

2. To the *difference in the colour of the seed albumen* [endosperm].* The albumen of the ripe seeds is either pale yellow,

* [Mendel uses the terms "albumen" and "endosperm" somewhat loosely to denote the cotyledons, containing food-material, within the seed.]

bright yellow and orange coloured, or it possesses a more or less intense green tint. This difference of colour is easily seen in the seeds as [= if] their coats are transparent.

3. To the *difference in the colour of the seed-coat.* This is either white, with which character white flowers are constantly correlated; or it is grey, grey-brown, leather-brown, with or without violet spotting, in which case the colour of the standards is violet, that of the wings purple, and the stem in the axils of the leaves is of a reddish tint. The grey seed-coats become dark brown in boiling water.

4. To the *difference in the form of the ripe pods.* These are either simply inflated, not contracted in places; or they are deeply constricted between the seeds and more or less wrinkled (*P. saccharatum*).

5. To the *difference in the colour of the unripe pods.* They are either light to dark green, or vividly yellow, in which colouring the stalks, leaf-veins, and calyx participate.*

6. To the *difference in the position of the flowers.* They are either axial, that is, distributed along the main stem; or they are terminal, that is, bunched at the top of the stem and arranged almost in a false umbel; in this case the upper part of the stem is more or less widened in section (*P. umbellatum*).†

7. To the *difference in the length of the stem.* The length of the stem is very various in some forms; it is, however, a constant character for each, in so far that healthy plants, grown in the same soil, are only subject to unimportant variations in this character.

In experiments with this character, in order to be able to discriminate with certainty, the long axis of 6 to 7 ft. was always crossed with the short one of $\frac{3}{4}$ ft. to $1\frac{1}{2}$ ft.

---

* One species possesses a beautifully brownish-red coloured pod, which when ripening turns to violet and blue. Trials with this character were only begun last year. [Of these further experiments it seems no account was published. Correns has since worked with such a variety.]

† [This is often called the Mummy Pea. It shows slight fasciation. The form I know has white standard and salmon-red wings.]

Each two of the differentiating characters enumerated above were united by cross-fertilisation. There were made for the

1st trial [[experiment]]	60	fertilisations	on	15 plants.
2nd ,,	58	,,	,,	10 ,,
3rd ,,	35	,,	,,	10 ,,
4th ,,	40	,,	,,	10 ,,
5th ,,	23	,,	,,	5 ,,
6th ,,	34	,,	,,	10 ,,
7th ,,	37	,,	,,	10 ,,

From a larger [[considerable]] number of plants of the same variety only the most vigorous were chosen for fertilisation. Weakly plants always afford uncertain results, because even in the first generation of hybrids, and still more so in the subsequent ones, many of the offspring either entirely fail to flower or only form a few and inferior seeds.

Furthermore, in all the experiments reciprocal crossings were effected in such a way that each of the two varieties which in one set of fertilisations served as seed-bearer in the other set was used as the pollen plant.

The plants were grown in garden beds, a few also in pots, and were maintained in their natural upright position by means of sticks, branches of trees, and strings stretched between. For each experiment a number of pot plants were placed during the blooming period in a greenhouse, to serve as control plants for the main experiment in the open as regards possible disturbance by insects. Among the insects * which visit Peas the beetle *Bruchus pisi* might be detrimental to the experiments should it appear in numbers. The female of this species is known to lay the eggs in the flower, and in so doing opens the keel; upon the tarsi of one specimen, which was caught in a flower, some pollen grains could clearly be seen under a lens. Mention must also be made of a circumstance which possibly might lead to the introduction of foreign pollen. It occurs, for instance, in some rare cases that certain parts of an otherwise quite normally developed flower wither, resulting in a partial exposure of the fertilising organs. A

* [It is somewhat surprising that no mention is made of Thrips, which swarm in Pea flowers. I had come to the conclusion that this is a real source of error and I see Laxton held the same opinion.]

defective development of the keel has also been observed, owing to which the stigma and anthers remained partially uncovered.* It also sometimes happens that the pollen does not reach full perfection. In this event there occurs a gradual lengthening of the pistil during the blooming period, until the stigmatic tip protrudes at the point of the keel. This remarkable appearance has also been observed in hybrids of *Phaseolus* and *Lathyrus*.

The risk of false impregnation by foreign pollen is, however, a very slight one with *Pisum*, and is quite incapable of disturbing the general result. Among more than 10,000 plants which were carefully examined there were only a very few cases where an indubitable false impregnation had occurred. Since in the greenhouse such a case was never remarked, it may well be supposed that *Bruchus pisi*, and possibly also the described abnormalities in the floral structure, were to blame.

4. [F_1] *THE FORMS OF THE HYBRIDS*†

Experiments which in previous years were made with ornamental plants have already afforded evidence that the hybrids, as a rule, are not exactly intermediate between the parental species. With some of the more striking characters, those, for instance, which relate to the form and size of the leaves, the pubescence of the several parts, etc., the intermediate, indeed, is nearly always to be seen; in other cases, however, one of the two parental characters is so preponderant that it is difficult, or quite impossible, to detect the other in the hybrid.

This is precisely the case with the Pea hybrids. In the case of each of the seven crosses the hybrid-character resembles ‡ that of one of the parental forms so closely that the other either escapes observation completely or cannot be detected with certainty. This circumstance is of great importance in

* [This also happens in Sweet Peas.]

† [Mendel throughout speaks of his cross-bred Peas as "hybrids", a term which many restrict to the offspring of two distinct *species*. He, as he explains, held this to be only a question of degree.]

‡ [Note that Mendel, with true penetration, avoids speaking of the hybrid-character as "transmitted" by either parent, thus escaping the error pervading the older views of heredity.]

the determination and classification of the forms under which the offspring of the hybrids appear. Henceforth in this paper those characters which are transmitted entire, or almost unchanged in the hybridisation, and therefore in themselves constitute the characters of the hybrid, are termed the *dominant*, and those which become latent in the process *recessive*. The expression "recessive" has been chosen because the characters thereby designated withdraw or entirely disappear in the hybrids, but nevertheless reappear unchanged in their progeny, as will be demonstrated later on.

It was furthermore shown by the whole of the experiments that it is perfectly immaterial whether the dominant character belongs to the seed plant or to the pollen plant; the form of the hybrid remains identical in both cases. This interesting fact was also emphasised by Gärtner, with the remark that even the most practised expert is not in a position to determine in a hybrid which of the two parental species was the seed or the pollen plant.*

Of the differentiating characters which were used in the experiments the following are dominant:

1. The round or roundish form of the seed with or without shallow depressions.

2. The yellow colouring of the seed albumen [cotyledons].

3. The grey, grey-brown, or leather-brown colour of the seed-coat, in association with violet-red blossoms and reddish spots in the leaf axils.

4. The simply inflated form of the pod.

5. The green colouring of the unripe pod in association with the same colour in the stems, the leaf-veins and the calyx.

6. The distribution of the flowers along the stem.

7. The greater length of stem.

With regard to this last character it must be stated that the longer of the two parental stems is usually exceeded by the hybrid, a fact which is possibly only attributable to the greater luxuriance which appears in all parts of plants when stems of very different length are crossed. Thus, for instance, in repeated

* [Gärtner, p. 223.]

experiments, stems of 1 ft. and 6 ft. in length yielded without exception hybrids which varied in length between 6 ft. and $7\frac{1}{2}$ ft.

The hybrid seeds in the experiments with seed-coat are often more spotted, and the spots sometimes coalesce into small bluish-violet patches. The spotting also frequently appears even when it is absent as a parental character.*

The hybrid forms of the seed-shape and of the albumen [colour] are developed immediately after the artificial fertilisation by the mere influence of the foreign pollen. They can, therefore, be observed even in the first year of experiment, whilst all the other characters naturally only appear in the following year in such plants as have been raised from the crossed seed.

5. [F_2] *THE FIRST GENERATION [BRED] FROM THE HYBRIDS*

In this generation there reappear, together with the dominant characters, also the recessive ones with their peculiarities fully developed, and this occurs in the definitely expressed average proportion of three to one, so that among each four plants of this generation three display the dominant character and one the recessive. This relates without exception to all the characters which were investigated in the experiments. The angular wrinkled form of the seed, the green colour of the albumen, the white colour of the seed-coats and the flowers, the constrictions of the pods, the yellow colour of the unripe pod, of the stalk, of the calyx, and of the leaf venation, the umbel-like form of the inflorescence, and the dwarfed stem, all reappear in the numerical proportion given, without any essential alteration. *Transitional forms were not observed in any experiment.*

Since the hybrids resulting from reciprocal crosses are formed alike and present no appreciable difference in their subsequent development, consequently the results [of the reciprocal crosses] can be reckoned together in each experiment. The relative numbers which were obtained for each pair of differentiating characters are as follows:

Expt. 1. Form of seed.—From 253 hybrids 7324 seeds

* [This refers to the coats of the seeds borne by F_1 plants.]

were obtained in the second trial [[experimental]] year. Among them were 5474 round or roundish ones and 1850 angular wrinkled ones. Therefrom the ratio 2·96 to 1 is deduced.

Expt. 2. Colour of albumen.—258 plants yielded 8023 seeds, 6022 yellow, and 2001 green; their ratio, therefore, is as 3·01 to 1.

In these two experiments each pod yielded usually both kinds of seed. In well-developed pods which contained on the average six to nine seeds, it often happened that all the seeds were round (Expt. 1) or all yellow (Expt. 2); on the other hand there were never observed more than five wrinkled or five green ones in one pod. It appears to make no difference whether the pods are developed early or later in the hybrid or whether they spring from the main axis or from a lateral one. In some few plants only a few seeds developed in the first formed pods, and these possessed exclusively one of the two characters, but in the subsequently developed pods the normal proportions were maintained nevertheless.

As in separate pods, so did the distribution of the characters vary in separate plants. By way of illustration the first ten individuals from both series of experiments may serve.

	Experiment 1		Experiment 2	
	Form of Seed		Colour of Albumen	
Plants	Round	Angular	Yellow	Green
1	45	12	25	11
2	27	8	32	7
3	24	7	14	5
4	19	10	70	27
5	32	11	24	13
6	26	6	20	6
7	88	24	32	13
8	22	10	44	9
9	28	6	50	14
10	25	7	44	18

As extremes in the distribution of the two seed characters in one plant, there were observed in Expt. 1 an instance of 43 round and only 2 angular, and another of 14 round and 15 angular seeds. In Expt. 2 there was a case of 32 yellow and only 1 green seed, but also one of 20 yellow and 19 green.

These two experiments are important for the determination of the average ratios, because with a smaller number of experimental plants they show that very considerable fluctuations may occur. In counting the seeds, also, especially in Expt. 2, some care is requisite, since in some of the seeds of many plants the green colour of the albumen is less developed, and at first may be easily overlooked. The cause of this partial disappearance of the green colouring has no connection with the hybrid-character of the plants, as it likewise occurs in the parental variety. This peculiarity [bleaching] is also confined to the individual and is not inherited by the offspring. In luxuriant plants this appearance was frequently noted. Seeds which are damaged by insects during their development often vary in colour and form, but with a little practice in sorting, errors are easily avoided. It is almost superfluous to mention that the pods must remain on the plants until they are thoroughly ripened and have become dried, since it is only then that the shape and colour of the seed are fully developed.

Expt. 3. Colour of the seed-coats.—Among 929 plants, 705 bore violet-red flowers and grey-brown seed-coats; 224 had white flowers and white seed-coats, giving the proportion 3·15 to 1.

Expt. 4. Form of pods.—Of 1181 plants 882 had them simply inflated, and in 299 they were constricted. Resulting ratio, 2·95 to 1.

Expt. 5. Colour of the unripe pods.—The number of trial ⟦experimental⟧ plants was 580, of which 428 had green pods and 152 yellow ones. Consequently these stand in the ratio 2·82 to 1.

Expt. 6. Position of flowers.—Among 858 cases 651 had inflorescences axial and 207 terminal. Ratio, 3·14 to 1.

Expt. 7. Length of stem.—Out of 1064 plants, in 787 cases the stem was long, and in 277 short. Hence a mutual ratio of 2·84 to 1. In this experiment the dwarfed plants were carefully lifted and transferred to a special bed. This precaution was necessary, as otherwise they would have perished through being overgrown by their tall relatives. Even in their quite young state they can be easily picked out by their compact growth and thick dark-green foliage.*

If now the results of the whole of the experiments be brought together, there is found, as between the number of forms with the dominant and recessive characters, an average ratio of 2·98 to 1, or 3 to 1.

The dominant character can have here a *double signification*—viz. that of a parental character, or a hybrid-character.† In which of the two significations it appears in each separate case can only be determined by the following generation. As a parental character it must pass over unchanged to the whole of the offspring; as a hybrid-character, on the other hand, it must maintain the same behaviour as in the first generation [F_2].

6. [F_3] *THE SECOND GENERATION [BRED] FROM THE HYBRIDS*

Those forms which in the first generation [F_2] exhibit the recessive character do not further vary in the second generation [F_3] as regards this character; they remain constant in their offspring.

It is otherwise with those which possess the dominant character in the first generation [bred from the hybrids]. Of these *two*-thirds yield offspring which display the dominant and recessive characters in the proportion of 3 to 1, and thereby show exactly the same ratio as the hybrid forms, while only *one*-third remains with the dominant character constant.

The separate experiments yielded the following results:

Expt. 1. Among 565 plants which were raised from round seeds of the first generation, 193 yielded round seeds only, and remained therefore constant in this character; 372, however, gave both round and wrinkled seeds, in the proportion of 3 to 1. The number of the hybrids, therefore, as compared with the constants is 1·93 to 1.

---

* [This is true also of the dwarf or "Cupid" Sweet Peas.]

† [This paragraph presents the view of the hybrid-character as something incidental to the hybrid, and not "transmitted" to it—a true and fundamental conception here expressed probably for the first time.]

Expt. 2. Of 519 plants which were raised from seeds whose albumen was of yellow colour in the first generation, 166 yielded exclusively yellow, while 353 yielded yellow and green seeds in the proportion of 3 to 1. There resulted, therefore, a division into hybrid and constant forms in the proportion of 2·13 to 1.

For each separate trial ⟦experiment⟧ in the following experiments 100 plants were selected which displayed the dominant character in the first generation, and in order to ascertain the significance of this, ten seeds of each were cultivated.

Expt. 3. The offspring of 36 plants yielded exclusively grey-brown seed-coats, while of the offspring of 64 plants some had grey-brown and some had white.

Expt. 4. The offspring of 29 plants had only simply inflated pods; of the offspring of 71, on the other hand, some had inflated and some constricted.

Expt. 5. The offspring of 40 plants had only green pods; of the offspring of 60 plants some had green, some yellow ones.

Expt. 6. The offspring of 33 plants had only axial flowers; of the offspring of 67, on the other hand, some had axial and some terminal flowers.

Expt. 7. The offspring of 28 plants inherited the long axis, and those of 72 plants some the long and some the short axis.

In each of these experiments a certain number of the plants came constant with the dominant character. For the determination of the proportion in which the separation of the forms with the constantly persistent character results, the two first experiments are of especial importance, since in these a larger ⟦considerable⟧ number of plants can be compared. The ratios 1·93 to 1 and 2·13 to 1 gave together almost exactly the average ratio of 2 to 1. The sixth experiment gave a quite concordant result; in the others the ratio varies more or less, as was only to be expected in view of the smaller number of 100 trial ⟦experimental⟧ plants. Experiment 5, which shows the greatest departure, was repeated, and then in lieu of the ratio of 60 and 40, that of 65 and 35 resulted. *The average ratio of 2 to 1 appears, therefore, as fixed with certainty.* It is therefore demonstrated that, of those forms which possess the dominant character in the first generation, two-thirds have the hybrid-character, while one-third remains constant with the dominant character.

The ratio of 3 to 1, in accordance with which the distribution of the dominant and recessive characters results in the first generation, resolves itself therefore in all experiments into the ratio of 2 : 1 : 1 if the dominant character be differentiated according to its significance as a hybrid-character or as a parental one. Since the members of the first generation [F_2] spring directly from the seed of the hybrids [F_1], *it is now clear that the hybrids form seeds having one or other of the two differentiating characters, and of these one-half develop again the hybrid form, while the other half yield plants which remain constant and receive the dominant or the recessive characters [respectively] in equal numbers.*

7. THE SUBSEQUENT GENERATIONS [BRED] FROM THE HYBRIDS

The proportions in which the descendants of the hybrids develop and split up in the first and second generations presumably hold good for all subsequent progeny. Experiments 1 and 2 have already been carried through six generations, 3 and 7 through five, and 4, 5, and 6 through four, these experiments being continued from the third generation with a small number of plants, and no departure from the rule has been perceptible. The offspring of the hybrids separated in each generation in the ratio of 2 : 1 : 1 into hybrids and constant forms.

If A be taken as denoting one of the two constant characters, for instance the dominant, a, the recessive, and Aa the hybrid form in which both are conjoined, the expression

$$A+2Aa+a$$

shows the terms in the series for the progeny of the hybrids of two differentiating characters.

The observation made by Gärtner, Kölreuter, and others, that hybrids are inclined to revert to the parental forms, is also confirmed by the experiments described. It is seen that the number of the hybrids which arise from one fertilisation, as compared with the number of forms which become constant, and their progeny from generation to generation, is continually diminishing, but that nevertheless they could not entirely disappear. If an average equality of fertility in all plants in all generations be assumed, and if, furthermore, each hybrid forms seed of which one-half yields hybrids again, while the other half is constant to both characters in equal proportions, the ratio of numbers for the offspring in each generation is seen by the following summary, in which A and a denote again the two parental characters, and Aa the hybrid forms. For brevity's sake it may be assumed that each plant in each generation furnishes only 4 seeds.

Generation	A	Aa	a	Ratios $A : Aa : a$
1	1	2	1	1 : 2 : 1
2	6	4	6	3 : 2 : 3
3	28	8	28	7 : 2 : 7
4	120	16	120	15 : 2 : 15
5	496	32	496	31 : 2 : 31
n				$2^n-1 : 2 : 2^n-1$

In the tenth generation, for instance, $2^n-1 = 1023$. There result, therefore, in each 2048 plants which arise in this generation 1023 with the constant dominant character, 1023 with the recessive character, and only two hybrids.

8. *THE OFFSPRING OF HYBRIDS IN WHICH SEVERAL DIFFERENTIATING CHARACTERS ARE ASSOCIATED*

In the experiments above described plants were used which differed only in one essential character.* The next task consisted in ascertaining whether the law of development

* [This statement of Mendel's in the light of present knowledge is open to some misconception. Though his work makes it evident that such varieties may exist, it is very unlikely that Mendel could have had seven pairs of varieties such that the members of each pair differed from each other in *only* one considerable character (*wesentliches Merkmal*). The point is probably of little theoretical or practical consequence, but a rather heavy stress is thrown on "*wesentlich*".]

discovered in these applied to each pair of differentiating characters when several diverse characters are united in the hybrid by crossing. As regards the form of the hybrids in these cases, the experiments showed throughout that this invariably more nearly approaches to that one of the two parental plants which possesses the greater number of dominant characters. If, for instance, the seed plant has a short stem, terminal white flowers, and simply inflated pods; the pollen plant, on the other hand, a long stem, violet-red flowers distributed along the stem, and constricted pods; the hybrid resembles the seed parent only in the form of the pod; in the other characters it agrees with the pollen parent. Should one of the two parental types possess only dominant characters, then the hybrid is scarcely or not at all distinguishable from it.

Two experiments were made with a considerable number of plants. In the first experiment the parental plants differed in the form of the seed and in the colour of the albumen; in the second in the form of the seed, in the colour of the albumen, and in the colour of the seed-coats. Experiments with seed characters give the result in the simplest and most certain way.

In order to facilitate study of the data in these experiments, the different characters of the seed plant will be indicated by *A*, *B*, *C*, those of the pollen plant by *a*, *b*, *c*, and the hybrid forms of the characters by *Aa*, *Bb*, and *Cc*.

Expt. 1.—*AB*, seed parents; *ab*, pollen parents;
A, form round; *a*, form wrinkled;
B, albumen yellow; *b*, albumen green.

The fertilised seeds appeared round and yellow like those of the seed parents. The plants raised therefrom yielded seeds of four sorts, which frequently presented themselves in one pod. In all, 556 seeds were yielded by 15 plants, and of these there were:

315 round and yellow,
101 wrinkled and yellow,
108 round and green,
32 wrinkled and green.

All were sown the following year. Eleven of the round yellow seeds did not yield plants, and three plants did not form seeds. Among the rest:

38	had round yellow seeds	*AB*
65	round yellow and green seeds	*ABb*
60	round yellow and wrinkled yellow seeds	*AaB*
138	round yellow and green, wrinkled yellow and green seeds	*AaBb*

From the wrinkled yellow seeds 96 resulting plants bore seed, of which:

28	had only wrinkled yellow seeds	*aB*
68	wrinkled yellow and green seeds	*aBb.*

From 108 round green seeds 102 resulting plants fruited, of which:

35	had only round green seeds	*Ab*
67	round and wrinkled green seeds	*Aab.*

The wrinkled green seeds yielded 30 plants which bore seeds all of like character; they remained constant *ab*.

The offspring of the hybrids appeared therefore under nine different forms, some of them in very unequal numbers. When these are collected and co-ordinated we find:

38 plants with the sign	*AB*
35 ,, ,, ,,	*Ab*
28 ,, ,, ,,	*aB*
30 ,, ,, ,,	*ab*
65 ,, ,, ,,	*ABb*
68 ,, ,, ,,	*aBb*
60 ,, ,, ,,	*AaB*
67 ,, ,, ,,	*Aab*
138 ,, ,, ,,	*AaBb.*

The whole of the forms may be classed into three essentially different groups. The first includes those with the signs *AB*, *Ab*, *aB*, and *ab*: they possess only constant characters and do not vary again in the next generation ⟦following generations⟧. Each of these forms is represented on the average thirty-three times. The second group includes the signs *ABb*, *aBb*, *AaB*, *Aab*:

these are constant in one character and hybrid in another, and vary in the next generation only as regards the hybrid-character. Each of these appears on an average sixty-five times. The form *AaBb* occurs 138 times: it is hybrid in both characters, and behaves exactly as do the hybrids from which it is derived.

If the numbers in which the forms belonging to these classes appear be compared, the ratios of 1, 2, 4 are unmistakably evident. The numbers 33, 65, 138 present very fair approximations to the ratio numbers of 33, 66, 132.

The developmental series consists, therefore, of nine classes, of which four appear therein always once and are constant in both characters; the forms *AB*, *ab*, resemble the parental forms, the two others present combinations between the conjoined characters *A*, *a*, *B*, *b*, which combinations are likewise possibly constant. Four classes appear always twice, and are constant in one character and hybrid in the other. One class appears four times, and is hybrid in both characters. Consequently the offspring of the hybrids, if two kinds of differentiating characters are combined therein, are represented by the expression

$$AB+Ab+aB+ab+2ABb+2aBb+2AaB+2Aab+4AaBb.$$

This expression is indisputably a combination series in which the two expressions for the characters *A* and *a*, *B* and *b* are combined. We arrive at the full number of the classes of the series by the combination of the expressions:

$$A+2Aa+a$$
$$B+2Bb+b.$$

Expt. 2.

ABC, seed parents;	*abc*, pollen parents;
A, form round;	*a*, form wrinkled;
B, albumen yellow;	*b*, albumen green;
C, seed-coat grey-brown;	*c*, seed-coat white.

This experiment was made in precisely the same way as the previous one. Among all the experiments it demanded the most time and trouble. From 24 hybrids 687 seeds were obtained in all: these were all either spotted, grey-brown or grey-green, round or wrinkled.* From these in the following year 639 plants fruited, and as further investigation showed, there were among them:

8 plants *ABC*	22 plants *ABCc*	45 plants *ABbCc*
14 „ *ABc*	17 „ *AbCc*	36 „ *aBbCc*
9 „ *AbC*	25 „ *aBCc*	38 „ *AaBCc*
11 „ *Abc*	20 „ *abCc*	40 „ *AabCc*
8 „ *aBC*	15 „ *ABbC*	49 „ *AaBbC*
10 „ *aBc*	18 „ *ABbc*	48 „ *AaBbc*
10 „ *abC*	19 „ *aBbC*	
7 „ *abc*	24 „ *aBbc*	
	14 „ *AaBC*	78 „ *AaBbCc*
	18 „ *AaBc*	
	20 „ *AabC*	
	16 „ *Aabc*	

The whole expression contains 27 terms. Of these 8 are constant in all characters, and each appears on the average 10 times; 12 are constant in two characters, and hybrid in the third; each appears on the average 19 times; 6 are constant in one character and hybrid in the other two; each appears on the average 43 times. One form appears 78 times and is hybrid in all of the characters. The ratios 10, 19, 43, 78 agree so closely with the ratios 10, 20, 40, 80, or 1, 2, 4, 8, that this last undoubtedly represents the true value.

The development of the hybrids when the original parents differ in three characters results therefore according to the following expression:

$$ABC + ABc + AbC + Abc + aBC + aBc + abC + abc + 2ABCc + 2AbCc + 2aBCc + 2abCc + 2ABbC + 2ABbc + 2aBbC + 2aBbc + 2AaBC + 2AaBc + 2AabC + 2Aabc + 4ABbCc + 4aBbCc + 4AaBCc + 4AabCc + 4AaBbC + 4AaBbc + 8AaBbCc.$$

Here also is involved a combination series in which the

* [Note that Mendel does not state the cotyledon-colour of the first crosses in this case; for as the coats were thick, it could not have been seen without opening or peeling the seeds.]

expressions for the characters A and a, B and b, C and c, are united. The expressions

$$A + 2Aa + a$$
$$B + 2Bb + b$$
$$C + 2Cc + c$$

give all the classes of the series. The constant combinations which occur therein agree with all combinations which are possible between the characters A, B, C, a, b, c; two thereof, ABC and abc, resemble the two original parental stocks.

In addition, further experiments were made with a smaller number of experimental plants in which the remaining characters by twos and threes were united as hybrids: all yielded approximately the same results. There is therefore no doubt that for the whole of the characters involved in the experiments the principle applies that *the offspring of the hybrids in which several essentially different characters are combined exhibit the terms of a series of combinations, in which the developmental series for each pair of differentiating characters are united.* It is demonstrated at the same time that *the relation of each pair of different characters in hybrid union is independent of the other differences in the two original parental stocks.*

If n represent the number of the differentiating characters in the two original stocks, 3^n gives the number of terms of the combination series, 4^n the number of individuals which belong to the series, and 2^n the number of unions which remain constant. The series therefore contains, if the original stocks differ in four characters, $3^4 = 81$ classes, $4^4 = 256$ individuals, and $2^4 = 16$ constant forms: or, which is the same, among each 256 offspring of the hybrids there are 81 different combinations, 16 of which are constant.

All constant combinations which in Peas are possible by the combination of the said seven differentiating characters were actually obtained by repeated crossing. Their number is given by $2^7 = 128$. Thereby is simultaneously given the practical proof *that the constant characters which appear in the several varieties of a group of plants may be obtained in all the associations which are possible according to the* [*mathematical*] *laws of combination, by means of repeated artificial fertilisation.*

As regards the flowering time of the hybrids, the experiments are not yet concluded. It can, however, already be stated that the time stands almost exactly between those of the seed and pollen parents, and that the constitution ⟦development⟧ of the hybrids with respect to this character probably follows the rule ascertained in the case of the other characters. The forms which are selected for experiments of this class must have a difference of at least twenty days from the middle flowering period of one to that of the other; furthermore, the seeds when sown must all be placed at the same depth in the earth, so that they may germinate simultaneously. Also, during the whole flowering period, the more important variations in temperature must be taken into account, and the partial hastening or delaying of the flowering which may result therefrom. It is clear that this experiment presents many difficulties to be overcome and necessitates great attention.

If we endeavour to collate in a brief form the results arrived at, we find that those differentiating characters, which admit of easy and certain recognition in the experimental plants, all behave exactly alike in their hybrid associations. The offspring of the hybrids of each pair of differentiating characters are, one-half, hybrid again, while the other half are constant in equal proportions having the characters of the seed and pollen parents respectively. If several differentiating characters are combined by cross-fertilisation in a hybrid, the resulting offspring form the terms of a combination series in which the combination ⟦developmental⟧ series for each pair of differentiating characters are united.

The uniformity of behaviour shown by the whole of the characters submitted to experiment permits, and fully justifies, the acceptance of the principle that a similar relation exists in the other characters which appear less sharply defined in plants, and therefore could not be included in the separate experiments. An experiment with peduncles of different lengths gave on the whole a fairly satisfactory result, although the differentiation and serial arrangement of the forms could not be effected with that certainty which is indispensable for correct experiment.

9. THE REPRODUCTIVE CELLS OF THE HYBRIDS

The results of the previously described experiments led to further experiments, the results of which appear fitted to afford some conclusions as regards the composition of the egg and pollen cells of hybrids. An important clue is afforded in *Pisum* by the circumstance that among the progeny of the hybrids constant forms appear, and that this occurs, too, in respect of all combinations of the associated characters. So far as experience goes, we find it in every case confirmed that constant progeny can only be formed when the egg cells and the fertilising pollen are of like character, so that both are provided with the material for creating quite similar individuals, as is the case with the normal fertilisation of pure species. We must therefore regard it as certain that exactly similar factors must be at work also in the production of the constant forms in the hybrid plants. Since the various constant forms are produced in *one* plant, or even in *one* flower of a plant, the conclusion appears logical that in the ovaries of the hybrids there are formed as many sorts of egg cells, and in the anthers as many sorts of pollen cells, as there are possible constant combination forms, and that these egg and pollen cells agree in their internal composition with those of the separate forms.

In point of fact it is possible to demonstrate theoretically that this hypothesis would fully suffice to account for the development of the hybrids in the separate generations, if we might at the same time assume that the various kinds of egg and pollen cells were formed in the hybrids on the average in equal numbers.*

In order to bring these assumptions to an experimental proof, the following experiments were designed. Two forms which were constantly different in the form of the seed and the colour of the albumen were united by fertilisation.

If the differentiating characters are again indicated as A, B, a, b, we have :

AB, seed parent;	ab, pollen parent;
A, form round;	a, form wrinkled;
B, albumen yellow;	b, albumen green.

* [This and the preceding paragraph contain the essence of the Mendelian principles of heredity.]

The artificially fertilised seeds were sown together with several seeds of both original stocks, and the most vigorous examples were chosen for the reciprocal crossing. There were fertilised:

1. The hybrids with the pollen of *AB*.
2. The hybrids ,, ,, *ab*.
3. *AB* ,, ,, the hybrids.
4. *ab* ,, ,, the hybrids.

For each of these four experiments the whole of the flowers on three plants were fertilised. If the above theory be correct, there must be developed on the hybrids egg and pollen cells of the forms *AB*, *Ab*, *aB*, *ab*, and there would be combined:

1. The egg cells *AB*, *Ab*, *aB*, *ab* with the pollen cells *AB*.
2. The egg cells *AB*, *Ab*, *aB*, *ab* with the pollen cells *ab*.
3. The egg cells *AB* with the pollen cells *AB*, *Ab*, *aB*, *ab*.
4. The egg cells *ab* with the pollen cells *AB*, *Ab*, *aB*, *ab*.

From each of these experiments there could then result only the following forms:

1. *AB*, *ABb*, *AaB*, *AaBb*.
2. *AaBb*, *Aab*, *aBb*, *ab*.
3. *AB*, *ABb*, *AaB*, *AaBb*.
4. *AaBb*, *Aab*, *aBb*, *ab*.

If, furthermore, the several forms of the egg and pollen cells of the hybrids were produced on an average in equal numbers, then in each experiment the said four combinations should stand in the same ratio to each other. A perfect agreement in the numerical relations was, however, not to be expected since in each fertilisation, even in normal cases, some egg cells remain undeveloped or subsequently die, and many even of the well-formed seeds fail to germinate when sown. The above assumption is also limited in so far that while it demands the formation of an equal number of the various sorts of egg and pollen cells, it does not require that this should apply to each separate hybrid with mathematical exactness.

The first and second experiments had primarily the object of proving the composition of the hybrid egg cells, while the third and fourth experiments were to decide that of the pollen cells.* As is shown by the above demonstration the first and third experiments and the second and fourth experiments should produce precisely the same combinations, and even in the second year the result should be partially visible in the form and colour of the artificially fertilised seed. In the first and third experiments the dominant characters of form and colour, *A* and *B*, appear in each union, and are also partly constant and partly in hybrid union with the recessive characters *a* and *b*, for which reason they must impress their peculiarity upon the whole of the seeds. All seeds should therefore appear round and yellow, if the theory be justified. In the second and fourth experiments, on the other hand, one union is hybrid in form and in colour, and consequently the seeds are round and yellow; another is hybrid in form, but constant in the recessive character of colour, whence the seeds are round and green; the third is constant in the recessive character of form but hybrid in colour, consequently the seeds are wrinkled and yellow; the fourth is constant in both recessive characters, so that the seeds are wrinkled and green. In both these experiments there were consequently four sorts of seed to be expected—viz. round and yellow, round and green, wrinkled and yellow, wrinkled and green.

The crop fulfilled these expectations perfectly. There were obtained in the

1st Experiment, 98 exclusively round yellow seeds;
3rd „ 94 „ „ „ „ .

In the 2nd Experiment, 31 round and yellow, 26 round and green, 27 wrinkled and yellow, 26 wrinkled and green seeds.

In the 4th Experiment, 24 round and yellow, 25 round and green, 22 wrinkled and yellow, 27 wrinkled and green seeds.

* [To prove, namely, that both were similarly differentiated, and not one or other only.]

There could scarcely be now any doubt of the success of the experiment; the next generation must afford the final proof. From the seed sown there resulted for the first experiment 90 plants, and for the third 87 plants which fruited: these yielded for the

1st Exp.	3rd Exp.		
20	25	round yellow seeds	*AB*
23	19	round yellow and green seeds . . .	*ABb*
25	22	round and wrinkled yellow seeds . .	*AaB*
22	21	round and wrinkled green and yellow seeds	*AaBb*

In the second and fourth experiments the round and yellow seeds yielded plants with round and wrinkled yellow and green seeds, *AaBb*.

From the round green seeds plants resulted with round and wrinkled green seeds, *Aab*.

The wrinkled yellow seeds gave plants with wrinkled yellow and green seeds, *aBb*.

From the wrinkled green seeds plants were raised which yielded again only wrinkled and green seeds, *ab*.

Although in these two experiments likewise some seeds did not germinate, the figures arrived at already in the previous year were not affected thereby, since each kind of seed gave plants which, as regards their seed, were like each other and different from the others. There resulted therefore from the

2nd. Exp.	4th. Exp.			
31	24	plants of the form		*AaBb*
26	25	,,	,,	*Aab*
27	22	,,	,,	*aBb*
26	27	,,	,,	*ab*

In all the experiments, therefore, there appeared all the forms which the proposed theory demands, and they came in nearly equal numbers.

In a further experiment the characters of flower-colour and length of stem were experimented upon, and selection was so made that in the third year of the experiment each character ought to appear in half of all the plants if the above theory were correct. *A*, *B*, *a*, *b* serve again as indicating the various characters.

A, violet-red flowers;	*a*, white flowers;
B, axis long;	*b*, axis short.

The form *Ab* was fertilised with *ab*, which produced the hybrid *Aab*. Furthermore, *aB* was also fertilised with *ab*, whence the hybrid *aBb*. In the second year, for further fertilisation, the hybrid *Aab* was used as seed parent, and hybrid *aBb* as pollen parent.

Seed parent, *Aab*;	Pollen parent, *aBb*;
Possible egg cells, *Ab*, *ab*;	Pollen cells, *aB*, *ab*.

From the fertilisation between the possible egg and pollen cells four combinations should result, viz.:

$$AaBb+aBb+Aab+ab.$$

From this it is perceived that, according to the above theory, in the third year of the experiment out of all the plants

half should have	violet-red flowers (*Aa*),	Classes	1, 3,
,, ,, ,,	white flowers (*a*)	,,	2, 4,
,, ,, ,,	a long axis (*Bb*)	,,	1, 2,
,, ,, ,,	a short axis (*b*)	,,	3, 4.

From 45 fertilisations of the second year 187 seeds resulted, of which only 166 reached the flowering stage in the third year. Among these the separate classes appeared in the numbers following:

Class	Colour of flower	Stem	
1	violet-red	long	47 times
2	white	long	40 ,,
3	violet-red	short	38 ,,
4	white	short	41 ,,

There subsequently [[consequently]] appeared

the violet-red flower-colour	(*Aa*)	in 85 plants,
the white flower-colour	(*a*)	in 81 plants,
the long stem	(*Bb*)	in 87 plants,
the short stem	(*b*)	in 79 plants.

The theory adduced is therefore satisfactorily confirmed in this experiment also.

For the characters of form of pod, colour of pod, and position of flowers, experiments were also made on a small scale and results obtained in perfect agreement. All combinations, which were possible through the union of the differentiating characters duly appeared, and in nearly equal numbers.

Experimentally, therefore, the theory is confirmed that *the pea hybrids form egg and pollen cells which, in their constitution, represent in equal numbers all constant forms which result from the combination of the characters united in fertilisation.*

The difference of the forms among the progeny of the hybrids, as well as the respective ratios of the numbers in which they are observed, find a sufficient explanation in the principle above deduced. The simplest case is afforded by the developmental series of each pair of differentiating characters. This series is represented by the expression $A+2Aa+a$, in which A and a signify the forms with constant differentiating characters, and Aa the hybrid form of both. It includes in three different classes four individuals. In the formation of these, pollen and egg cells of the form A and a take part on the average equally in the fertilisation; hence each form [occurs] twice, since four individuals are formed. There participate consequently in the fertilisation

the pollen cells $A+A+a+a$,
the egg cells $A+A+a+a$.

It remains, therefore, purely a matter of chance which of the two sorts of pollen will become united with each separate egg cell. According, however, to the law of probability, it will always happen, on the average of many cases, that each pollen form A and a will unite equally often with each egg cell form A and a, consequently one of the two pollen cells A in the fertilisation will meet with the egg cell A and the other with an egg cell a, and so likewise one pollen cell a will unite with an egg cell A, and the other with egg cell a.

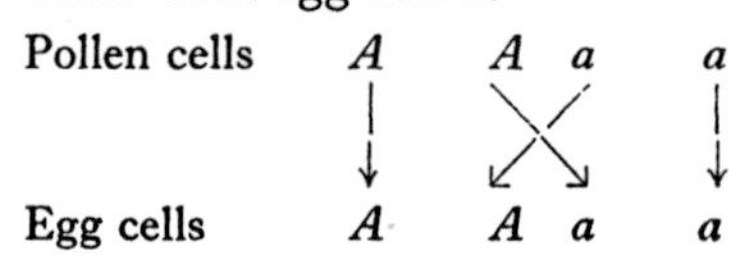

The result of the fertilisation may be made clear by putting the signs for the conjoined egg and pollen cells in the form of fractions, those for the pollen cells above and those for the egg cells below the line. We then have

$$\frac{A}{A} + \frac{A}{a} + \frac{a}{A} + \frac{a}{a}.$$

In the first and fourth term the egg and pollen cells are of like kind, consequently the product of their union must be constant, viz. A and a; in the second and third, on the other hand, there again results a union of the two differentiating characters of the stocks, consequently the forms resulting from these fertilisations are identical with those of the hybrid from which they sprang. *There occurs accordingly a repeated hybridisation.* This explains the striking fact that the hybrids are able to produce, besides the two parental forms, offspring which are like themselves; $\frac{A}{a}$ and $\frac{a}{A}$ both give the same union Aa, since, as already remarked above, it makes no difference in the result of fertilisation to which of the two characters the pollen or egg cells belong. We may write then

$$\frac{A}{A} + \frac{A}{a} + \frac{a}{A} + \frac{a}{a} = A + 2Aa + a.$$

This represents the average result of the self-fertilisation of the hybrids when two differentiating characters are united in them. In individual flowers and in individual plants, however, the ratios in which the forms of the series are produced may suffer not inconsiderable fluctuations.* Apart from the fact that the numbers in which both sorts of egg cells occur in the seed vessels can only be regarded as equal on the average, it remains purely a matter of chance which of the two sorts of pollen may fertilise each separate egg cell. For this reason the separate values must necessarily be subject to fluctuations, and there are even extreme cases possible, as were described

* [Whether segregation by such units is more than purely fortuitous may perhaps be determined by seriation.]

earlier in connection with the experiments on the form of the seed and the colour of the albumen. The true ratios of the numbers can only be ascertained by an average deduced from the sum of as many single values as possible; the greater the number the more are merely chance effects eliminated.

The developmental series for hybrids in which two kinds of differentiating characters are united contains among sixteen individuals nine different forms, viz.:

$$AB+Ab+aB+ab+2ABb+2aBb+2AaB+2Aab+4AaBb.$$

Between the differentiating characters of the original stocks *Aa* and *Bb* four constant combinations are possible, and consequently the hybrids produce the corresponding four forms of egg and pollen cells *AB*, *Ab*, *aB*, *ab*, and each of these will on the average figure four times in the fertilisation, since sixteen individuals are included in the series. Therefore the participators in the fertilisation are

Pollen cells $AB+AB+AB+AB+Ab+Ab+Ab+Ab+aB+aB+aB+aB+aB+ab+ab+ab+ab.$

Egg cells $AB+AB+AB+AB+Ab+Ab+Ab+Ab+aB+aB+aB+aB+aB+ab+ab+ab+ab.$

In the process of fertilisation each pollen form unites on an average equally often with each egg cell form, so that each of the four pollen cells *AB* unites once with one of the forms of egg cell *AB*, *Ab*, *aB*, *ab*. In precisely the same way the rest of the pollen cells of the forms *Ab*, *aB*, *ab* unite with all the other egg cells. We obtain therefore

$$\frac{AB}{AB}+\frac{AB}{Ab}+\frac{AB}{aB}+\frac{AB}{ab}+\frac{Ab}{AB}+\frac{Ab}{Ab}+\frac{Ab}{aB}+\frac{Ab}{ab}$$
$$+\frac{aB}{AB}+\frac{aB}{Ab}+\frac{aB}{aB}+\frac{aB}{ab}+\frac{ab}{AB}+\frac{ab}{Ab}+\frac{ab}{aB}+\frac{ab}{ab},$$

or

$$AB + ABb + AaB + AaBb + ABb + Ab + AaBb + Aab + AaB + AaBb + aB + aBb + AaBb + Aab + aBb + ab = AB + Ab + aB + ab + 2ABb + 2aBb + 2AaB + 2Aab + 4AaBb.$$

* [In the original the sign of equality (=) is here represented by +, evidently a misprint.]

In precisely similar fashion is the developmental series of hybrids exhibited when three kinds of differentiating characters are conjoined in them. The hybrids form eight various kinds of egg and pollen cells—*ABC*, *ABc*, *AbC*, *Abc*, *aBC*, *aBc*, *abC*, *abc*—and each pollen form unites itself again on the average once with each form of egg cell.

The law of combination of different characters which governs the development of the hybrids finds therefore its foundation and explanation in the principle enunciated, that the hybrids produce egg cells and pollen cells which in equal numbers represent all constant forms which result from the combinations of the characters brought together in fertilisation.

10. *EXPERIMENTS WITH HYBRIDS OF OTHER SPECIES OF PLANTS*

It must be the object of further experiments to ascertain whether the law of development discovered for *Pisum* applies also to the hybrids of other plants. To this end several experiments were recently commenced. Two minor experiments with species of *Phaseolus* have been completed, and may be here mentioned.

An experiment with *Phaseolus vulgaris* and *Phaseolus nanus* gave results in perfect agreement. *Ph. nanus* had together with the dwarf axis, simply inflated, green pods. *Ph. vulgaris* had, on the other hand, an axis 10 ft. to 12 ft. high, and yellow-coloured pods, constricted when ripe. The ratios of the numbers in which the different forms appeared in the separate generations were the same as with *Pisum*. Also the development of the constant combinations resulted according to the law of simple combination of characters, exactly as in the case of *Pisum*. There were obtained

Constant combinations	Axis	Colour of the unripe pods	Form of the ripe pods
1	long	green	inflated
2	,,	,,	constricted
3	,,	yellow	inflated
4	,,	,,	constricted
5	short	green	inflated
6	,,	,,	constricted
7	,,	yellow	inflated
8	,,	,,	constricted

The green colour of the pod, the inflated forms, and the long axis were, as in *Pisum*, dominant characters.

Another experiment with two very different species of *Phaseolus* had only a partial result. *Phaseolus nanus*, L., served as seed parent, a perfectly constant species, with white flowers in short racemes and small white seeds in straight, inflated, smooth pods; as pollen parent was used *Ph. multiflorus*, W., with tall winding stem, purple-red flowers in very long racemes, rough, sickle-shaped crooked pods, and large seeds which bore black flecks and splashes on a peach-blood-red ground.

The hybrids had the greatest similarity to the pollen parent, but the flowers appeared less intensely coloured. Their fertility was very limited; from seventeen plants, which together developed many hundreds of flowers, only forty-nine seeds in all were obtained. These were of medium size, and were flecked and splashed similarly to those of *Ph. multiflorus*, while the ground colour was not materially different. The next year forty-four plants were raised from these seeds, of which only thirty-one reached the flowering stage. The characters of *Ph. nanus*, which had been altogether latent in the hybrids, reappeared in various combinations; their ratio, however, with relation to the dominant plants was necessarily very fluctuating owing to the small number of trial plants. With certain characters, as in those of the axis and the form of pod, it was, however, as in the case of *Pisum*, almost exactly 1 : 3.

Insignificant as the results of this experiment may be as regards the determination of the relative numbers in which the various forms appeared, it presents, on the other hand, the phenomenon of a remarkable change of colour in the flowers and seed of the hybrids. In *Pisum* it is known that the characters of the flower- and seed-colour present themselves unchanged in the first and further generations, and that the offspring of the hybrids display exclusively the one or the other of the characters of the original stocks. It is otherwise in the experiment we are considering. The white flowers and the seed-colour of *Ph. nanus* appeared, it is true, at once in the first generation [*from* the hybrids] in one fairly fertile example, but the remaining thirty plants developed flower-colours which were of various grades of purple-red to pale violet. The colouring of the seed-coat was no less varied than that of the flowers. No plant could rank as fully fertile; many produced no fruit at all; others only yielded fruits from the flowers last produced, which did not ripen. From fifteen plants only were well-developed seeds obtained. The greatest disposition to infertility was seen in the forms with preponderantly red flowers, since out of sixteen of these only four yielded ripe seed. Three of these had a similar seed pattern to *Ph. multiflorus*, but with a more or less pale ground colour; the fourth plant yielded only one seed of plain brown tint. The forms with preponderantly violet-coloured flowers had dark brown, black-brown, and quite black seeds.

The experiment was continued through two more generations under similar unfavourable circumstances, since even among the offspring of fairly fertile plants there came again some which were less fertile or even quite sterile. Other flower- and seed-colours than those cited did not subsequently present themselves. The forms which in the first generation [bred from the hybrids] contained one or more of the recessive characters remained, as regards these, constant without exception. Also of those plants which possessed violet flowers and brown or black seed, some did not vary again in these respects in the next generation⟦s⟧; the majority, however, yielded, together with offspring exactly like themselves, some which displayed white flowers and white seed-coats. The red flowering plants remained so slightly fertile that nothing can be said with certainty as regards their further development.

Despite the many disturbing factors with which the observations had to contend, it is nevertheless seen by this experiment that the development of the hybrids, with regard to those characters which concern the form of the plants, follows the same laws as in *Pisum*. With regard to the colour characters, it certainly appears difficult to perceive a substantial agreement. Apart from the fact that from the union of a white and a purple-red colouring a whole series of colours results [in F_2], from purple to pale violet and white, the circumstance is a striking one that among thirty-one flowering plants only one received the recessive character of the white colour, while in *Pisum* this occurs on the average in every fourth plant.

Even these enigmatical results, however, might probably be explained by the law governing *Pisum* if we might assume that the colour of the flowers and seeds of *Ph. multiflorus* is a combination of two or more entirely independent colours, which individually act like any other constant character in the plant. If the flower-colour A were a combination of the individual characters $A_1+A_2+\dots$ which produce the total impression of a purple coloration, then by fertilisation with the differentiating character, white colour, a, there would be produced the hybrid unions $A_1a+A_2a+\dots$ and so would it be with the corresponding colouring of the seed-coats.* According to the above assumption, each of these hybrid colour unions would be independent, and would consequently develop quite independently from the others. It is then easily seen that from the combination of the separate developmental series a complete colour-series must result. If, for instance, $A = A_1+A_2$, then the hybrids A_1a and A_2a form the developmental series—

$$A_1+2A_1a+a$$
$$A_2+2A_2a+a.$$

The members of this series can enter into nine different combinations, and each of these denotes another colour—

1 A_1A_2	2 A_1aA_2	1 A_2a
2 A_1A_2a	4 A_1aA_2a	2 A_2aa
1 A_1a	2 A_1aa	1 aa.

The figures prescribed for the separate combinations also indicate how many plants with the corresponding colouring belong to the series. Since the total is sixteen, the whole of the colours are on the average distributed over each sixteen plants, but, as the series itself indicates, in unequal proportions.

Should the colour development really happen in this way, we could offer an explanation of the case above described, viz. that the white flowers and seed-coat colour only appeared

* [As it fails to take account of factors introduced by the albino this representation is imperfect. It is however interesting to know that Mendel realised the fact of the existence of compound characters, and that the rarity of the white recessives was a consequence of this resolution.]

once among thirty-one plants of the first generation. This colouring appears only once in the series, and could therefore also only be developed once in the average in each sixteen, and with three colour characters only once even in sixty-four plants.

It must, nevertheless, not be forgotten that the explanation here attempted is based on a mere hypothesis, only supported by the very imperfect result of the experiment just described. It would, however, be well worth while to follow up the development of colour in hybrids by similar experiments, since it is probable that in this way we might learn the significance of the extraordinary variety in the colouring of our ornamental flowers.

So far, little at present is known with certainty beyond the fact that the colour of the flowers in most ornamental plants is an extremely variable character. The opinion has often been expressed that the stability of the species is greatly disturbed or entirely upset by cultivation, and consequently there is an inclination to regard the development of cultivated forms as a matter of chance devoid of rules; the colouring of ornamental plants is indeed usually cited as an example of great instability. It is, however, not clear why the simple transference into garden soil should result in such a thorough and persistent revolution in the plant organism. No one will seriously maintain that in the open country the development of plants is ruled by other laws than in the garden bed. Here, as there, changes of type must take place if the conditions of life be altered, and the species possesses the capacity of fitting itself to its new environment. It is willingly granted that by cultivation the origination of new varieties is favoured, and that by man's labour many varieties are acquired which, under natural conditions, would be lost; but nothing justifies the assumption that the tendency to the formation of varieties is so extraordinarily increased that the species speedily lose all stability, and their offspring diverge into an endless series of extremely variable forms. Were the change in the conditions the sole cause of variability we might expect that those cultivated plants which are grown for centuries under almost identical conditions would again attain constancy. That, as is well known, is not the case since it is precisely under such circumstances that not only the most varied but also the most variable forms are found. It is only the *Leguminosae*, like *Pisum*, *Phaseolus*,* *Lens*, whose organs of fertilisation are protected by the keel, which constitute a noteworthy exception. Even here there have arisen numerous varieties during a cultural period of more than 1000 years under most various conditions; these maintain, however, under unchanging environments a stability as great as that of species growing wild.

It is more than probable that as regards the variability of cultivated plants there exists a factor which so far has received little attention. Various experiments force us to the conclusion that our cultivated plants, with few exceptions, are *members of various hybrid series*, whose further development in conformity with law is varied and interrupted by frequent crossings *inter se*. The circumstance must not be overlooked that cultivated plants are mostly grown in great numbers and close together, affording the most favourable conditions for reciprocal fertilisation between the varieties present and the species itself. The probability of this is supported by the fact that among the great array of variable forms solitary examples are always found, which in one character or another remain constant, if only foreign influence be carefully excluded. These forms behave precisely as do those which are known to be members of the compound hybrid series. Also with the most susceptible of all characters, that of colour, it cannot escape the careful observer that in the separate forms the inclination to vary is displayed in very different degrees. Among plants which arise from *one* spontaneous fertilisation there are often some whose offspring vary widely in the constitution and arrangement of the colours, while that of others shows little deviation, and among a greater number solitary examples occur which transmit the colour of the flowers unchanged to their offspring. The cultivated species of *Dianthus* afford an instructive example of this. A white-flowered example of *Dianthus caryophyllus*, which itself was derived from a white-flowered variety, was shut up during its blooming period in a

* [*Phaseolus* nevertheless is insect-fertilised.]

greenhouse; the numerous seeds obtained therefrom yielded plants entirely white-flowered like itself. A similar result was obtained from a sub-species, with red flowers somewhat flushed with violet, and one with flowers white, striped with red. Many others, on the other hand, which were similarly protected, yielded progeny which were more or less variously coloured and marked.

Whoever studies the coloration which results in ornamental plants from similar fertilisation can hardly escape the conviction that here also the development follows a definite law which possibly finds its expression *in the combination of several independent colour characters.*

11. *CONCLUDING REMARKS*

It can hardly fail to be of interest to compare the observations made regarding *Pisum* with the results arrived at by the two authorities in this branch of knowledge, Kölreuter and Gärtner, in their investigations. According to the opinion of both, the hybrids in outward appearance present either a form intermediate between the original species, or they closely resemble either the one or the other type, and sometimes can hardly be discriminated from it. From their seeds usually arise, if the fertilisation was effected by their own pollen, various forms which differ from the normal type. As a rule, the majority of individuals obtained by one fertilisation maintain the hybrid form, while some few others come more like the seed parent, and one or other individual approaches the pollen parent. This, however, is not the case with all hybrids without exception. Sometimes the offspring have more nearly approached, some the one and some the other of the two original stocks, or they all incline more to one or the other side; while in other cases *they remain perfectly like the hybrid* and continue constant in their offspring. The hybrids of varieties behave like hybrids of species, but they possess greater variability of form and a more pronounced tendency to revert to the original types.

With regard to the form of the hybrids and their development, as a rule an agreement with the observations made in

Pisum is unmistakable. It is otherwise with the exceptional cases cited. Gärtner confesses even that the exact determination whether a form bears a greater resemblance to one or to the other of the two original species often involved great difficulty, so much depending upon the subjective point of view of the observer. Another circumstance could, however, contribute to render the results fluctuating and uncertain, despite the most careful observation and differentiation. For the experiments, plants were mostly used which rank as good species and are differentiated by a large number of characters. In addition to the sharply defined characters, where it is a question of greater or less similarity, those characters must also be taken into account which are often difficult to define in words, but yet suffice, as every plant specialist knows, to give the forms a peculiar appearance. If it be accepted that the development of hybrids follows the law which is valid for *Pisum*, the series in each separate experiment must contain very many forms, since the number of the terms, as is known, increases with the number of the differentiating characters as the powers of three. With a relatively small number of experimental plants the result therefore could only be approximately right, and in single cases might fluctuate considerably. If, for instance, the two original stocks differ in seven characters, and 100-200 plants were raised from the seeds of their hybrids to determine the grade of relationship of the offspring, we can easily see how uncertain the decision must become since for seven differentiating characters the combination ⟦developmental⟧ series contains 16,384 individuals under 2187 various forms; now one and then another relationship could assert its predominance, just according as chance presented this or that form to the observer in a majority of cases.

If, furthermore, there appear among the differentiating characters at the same time *dominant* characters, which are transmitted entire or nearly unchanged to the hybrids, then in the terms of the developmental series that one of the two original parents which possesses the majority of dominant characters must always be predominant. In the experiment described relative to *Pisum*, in which three kinds of differentiating characters were concerned, all the dominant characters belonged to the seed parent. Although the terms of the series in their internal composition approach both original parents equally, yet in this experiment the type of the seed parent obtained so great a preponderance that out of each sixty-four plants of the first generation fifty-four exactly resembled it, or only differed in one character. It is seen how rash it must be under such circumstances to draw from the external resemblances of hybrids conclusions as to their internal nature.

Gärtner mentions that in those cases where the development was regular among the offspring of the hybrids the two original species were not reproduced, but only a few individuals which approached them. With very extended developmental series it could not in fact be otherwise. For seven differentiating characters, for instance, among more than 16,000 individuals—offspring of the hybrids—each of the two original species would occur only once. It is therefore hardly possible that these should appear at all among a small number of experimental plants; with some probability, however, we might reckon upon the appearance in the series of a few forms which approach them.

We meet with an *essential difference* in those hybrids which remain constant in their progeny and propagate themselves as truly as the pure species. According to Gärtner, to this class belong the *remarkably fertile hybrids Aquilegia atropurpurea canadensis*, *Lavatera pseudolbia thuringiaca*, *Geum urbano-rivale*, and some *Dianthus* hybrids; and, according to Wichura, the hybrids of the Willow family. For the history of the evolution of plants this circumstance is of special importance, since constant hybrids acquire the status of new species. The correctness of the facts is guaranteed by eminent observers, and cannot be doubted. Gärtner had an opportunity of following up *Dianthus Armeria deltoides* to the tenth generation, since it regularly propagated itself in the garden.

With *Pisum* it was shown by experiment that the hybrids form egg and pollen cells of *different* kinds, and that herein lies the reason of the variability of their offspring. In other hybrids, likewise, whose offspring behave similarly we may assume a like cause; for those, on the other hand, which remain constant the assumption appears justifiable that their reproductive cells are all alike and agree with the foundation-cell [fertilised ovum] of the hybrid. In the opinion of renowned physiologists, for the purpose of propagation one pollen cell and one egg cell unite in Phanerogams * into a single cell, which is capable by assimilation and formation of new cells to become an independent organism. This development follows a constant law, which is founded on the material composition and arrangement of the elements which meet in the cell in a vivifying union. If the reproductive cells be of the same kind and agree with the foundation cell [fertilised ovum] of the mother plant, then the development of the new individual will follow the same law which rules the mother plant. If it chance that an egg cell unites with a *dissimilar* pollen cell, we must then assume that between those elements of both cells, which determine opposite characters some sort of compromise is effected. The resulting compound cell becomes the foundation of the hybrid organism the development of which necessarily follows a different scheme from that obtaining in each of the two original species. If the compromise be taken to be a complete one, in the sense, namely, that the hybrid embryo is formed from two similar cells, in which the differences are *entirely and permanently accommodated* together, the further result follows that the hybrids, like any other stable plant species, reproduce themselves truly in their offspring. The reproductive cells which are formed in their seed vessels and anthers are of one kind, and agree with the fundamental compound cell [fertilised ovum].

With regard to those hybrids whose progeny is *variable* we may perhaps assume that between the differentiating elements of the egg and pollen cells there also occurs a compromise, in so far that the formation of a cell as foundation of

---

* In *Pisum* it is placed beyond doubt that for the formation of the new embryo a perfect union of the elements of both reproductive cells must take place. How could we otherwise explain that among the offspring of the hybrids both original types reappear in equal numbers and with all their peculiarities? If the influence of the egg cell upon the pollen cell were only external, if it fulfilled the *rôle* of a nurse only, then the result of each artificial fertilisation could be no other than that the developed hybrid should exactly resemble the pollen parent, or at any rate do so very closely. This the experiments so far have in no wise confirmed. An evident proof of the complete union of the contents of both cells is afforded by the experience gained on all sides that it is immaterial, as regards the form of the hybrid, which of the original species is the seed parent or which the pollen parent.

the hybrid becomes possible; but, nevertheless, the arrangement between the conflicting elements is only temporary and does not endure throughout the life of the hybrid plant. Since in the habit of the plant no changes are perceptible during the whole period of vegetation, we must further assume that it is only possible for the differentiating elements to liberate themselves from the enforced union when the fertilising cells are developed. In the formation of these cells all existing elements participate in an entirely free and equal arrangement, by which it is only the differentiating ones which mutually separate themselves. In this way the production would be rendered possible of as many sorts of egg and pollen cells as there are combinations possible of the formative elements.

The attribution attempted here of the essential difference in the development of hybrids to *a permanent or temporary union* of the differing cell elements can, of course, only claim the value of an hypothesis for which the lack of definite data offers a wide scope. Some justification of the opinion expressed lies in the evidence afforded by *Pisum* that the behaviour of each pair of differentiating characters in hybrid union is independent of the other differences between the two original plants, and, further, that the hybrid produces just so many kinds of egg and pollen cells as there are possible constant combination forms. The differentiating characters of two plants can finally, however, only depend upon differences in the composition and grouping of the elements which exist in the foundation-cells [fertilised ova] of the same in vital interaction.*

Even the validity of the law formulated for *Pisum* requires still to be confirmed, and a repetition of the more important experiments is consequently much to be desired, that, for instance, relating to the composition of the hybrid fertilising cells. A differential [element] may easily escape the single observer,† which although at the outset may appear to be unimportant, may yet accumulate to such an extent that it must not be ignored in the total result. Whether the variable hybrids of other plant species observe an entire agreement must also be

* ["Welche inden Grundzellen derselben in lebendiger Wechselwirkung stehen."]

† ["Dem einzelnen Beobachter kann leicht ein Differenziale entgehen."]

first decided experimentally. In the meantime we may assume that in material points an essential difference can scarcely occur, since the unity in the developmental plan of organic life is beyond question.

In conclusion, the experiments carried out by Kölreuter, Gärtner, and others with respect to *the transformation of one species into another by artificial fertilisation* merit special mention. Particular importance has been attached to these experiments, and Gärtner reckons them among "the most difficult of all in hybridisation".

If a species *A* is to be transformed into a species *B*, both must be united by fertilisation and the resulting hybrids then be fertilised with the pollen of *B*; then, out of the various offspring resulting, that form would be selected which stood in nearest relation to *B* and once more be fertilised with *B* pollen, and so continuously until finally a form is arrived at which is like *B* and constant in its progeny. By this process the species *A* would change into the species *B*. Gärtner alone has effected thirty such experiments with plants of genera *Aquilegia*, *Dianthus*, *Geum*, *Lavatera*, *Lychnis*, *Malva*, *Nicotiana*, and *Oencthera*. The period of transformation was not alike for all species. While with some a triple fertilisation sufficed, with others this had to be repeated five or six times, and even in the same species fluctuations were observed in various experiments. Gärtner ascribes this difference to the circumstance that "the specific [*typische*] power by which a species, during reproduction, effects the change and transformation of the maternal type varies considerably in different plants, and that, consequently, the periods within which the one species is changed into the other must also vary, as also the number of generations, so that the transformation in some species is perfected in more, and in others in fewer generations". Further, the same observer remarks "that in these transformation experiments a good deal depends upon which type and which individual be chosen for further transformation".

If it may be assumed that in these experiments the constitution of the forms resulted in a similar way to that of *Pisum*, the entire process of transformation would find a fairly simple explanation. The hybrid forms as many kinds of egg cells as there are constant combinations possible of the characters conjoined therein, and one of these is always of the same kind as that of the fertilising pollen cells. Consequently there always exists the possibility with all such experiments that even from the second fertilisation there may result a constant form identical with that of the pollen parent. Whether this really be obtained depends in each separate case upon the number of the experimental plants, as well as upon the number of differentiating characters which are united by the fertilisation. Let us, for instance, assume that the plants selected for experiment differed in three characters, and the species *ABC* is to be transformed into the other species *abc* by repeated fertilisation with the pollen of the latter; the hybrids resulting from the first cross form eight different kinds of egg cells, viz.:

$$ABC,\ ABc,\ AbC,\ aBC,\ Abc,\ aBc,\ abC,\ abc.$$

These in the second year of experiment are united again with the pollen cells *abc*, and we obtain the series

$$AaBbCc + AaBbc + AabCc + aBbCc + Aabc + aBbc + abCc + abc.$$

Since the form *abc* occurs once in the series of eight terms, it is consequently little likely that it would be missing among the experimental plants, even were these raised in a smaller number, and the transformation would be perfected already by a second fertilisation. If by chance it did not appear, then the fertilisation must be repeated with one of those forms nearest akin, *Aabc*, *aBbc*, *abCc*. It is perceived that such an experiment must extend the farther *the smaller the number of experimental plants and the larger the number of differentiating characters* in the two original species; and that, furthermore, in the same species there can easily occur a delay of one or even of two generations such as Gärtner observed. The transformation of widely divergent species could generally only be completed in five or six years of experiment, since the number of different egg cells which are formed in the hybrid increases as the powers of two with the number of differentiating characters.

Gärtner found by repeated experiments that the respective period of transformation varies in many species, so that frequently a species *A* can be transformed into a species *B* a generation sooner than can species B into species A. He deduces therefrom that Kölreuter's opinion can hardly be maintained that "the two natures in hybrids are perfectly in equilibrium". It appears, however, that Kölreuter does not merit this criticism, but that Gärtner rather has overlooked a material point, to which he himself elsewhere draws attention, viz. that "it depends which individual is chosen for further transformation". Experiments which in this connection were carried out with two species of *Pisum* demonstrated that as regards the choice of the fittest individuals for the purpose of further fertilisation it may make a great difference which of two species is transformed into the other. The two experimental plants differed in five characters, while at the same time those of species A were all dominant and those of species B all recessive. For mutual transformation A was fertilised with pollen of B, and B with pollen of A, and this was repeated with both hybrids the following year. With the first experiment $\frac{B}{A}$ there were eighty-seven plants available in the third year of experiment for selection of the individuals for further crossing, and these were of the possible thirty-two forms; with the second experiment $\frac{A}{B}$ seventy-three plants resulted, which *agreed throughout perfectly in habit with the pollen parent*; in their internal composition, however, they must have been just as varied as the forms in the other experiment. A definite selection was consequently only possible with the first experiment; with the second the selection had to be made at random, merely. Of the latter only a portion of the flowers were crossed with the A pollen, the others were left to fertilise themselves. Among each five plants which were selected in both experiments for fertilisation there agreed, as the following year's culture showed, with the pollen parent:

1st Experiment	2nd Experiment	
2 plants	—	in all characters
3 ,,	—	,, 4 ,,
—	2 plants	,, 3 ,,
—	2 ,,	,, 2 ,,
—	1 plant	,, 1 character

In the first experiment, therefore, the transformation was completed; in the second, which was not continued further, two more fertilisations would probably have been required.

Although the case may not frequently occur in which the dominant characters belong exclusively to one or the other of the original parent plants, it will always make a difference which of the two possesses the majority of dominants. If the pollen parent has the majority, then the selection of forms for further crossing will afford a less degree of certainty than in the reverse case, which must imply a delay in the period of transformation, provided that the experiment is only considered as completed when a form is arrived at which not only exactly resembles the pollen plant in form, but also remains as constant in its progeny.

Gärtner, by the results of these transformation experiments, was led to oppose the opinion of those naturalists who dispute the stability of plant species and believe in a continuous evolution of vegetation. He perceives* in the complete transformation of one species into another an indubitable proof that species are fixed within limits beyond which they cannot change. Although this opinion cannot be unconditionally accepted we find on the other hand in Gärtner's experiments a noteworthy confirmation of that supposition regarding variability of cultivated plants which has already been expressed.

Among the experimental species there were cultivated plants, such as *Aquilegia atropurpurea* and *canadensis*, *Dianthus caryophyllus*, *chinensis*, and *japonicus*, *Nicotiana rustica* and *paniculata*, and hybrids between these species lost none of their stability after four or five generations.

---

* ["Es sieht" in the original is clearly a misprint for "Er sieht".]

CPSIA information can be obtained at www.ICGtesting.com
Printed in the USA
LVOW101645201112

308200LV00002B/54/P